ELEMENTARY SURVEYING

FOR INDUSTRIAL ARCHAEOLOGISTS

D0514499

HUGH BODEY
and
MICHAEL HALLAS

SHIRE PUBLICATIONS LTD

Printed in Great Britain by C. I. Thomas & Sons (Haverfordwest) Ltd, Press Buildings, Merlins Bridge, Haverfordwest, Dyfed.

ELEVATION

PLAN AT GROUND FLOOR.

Contents

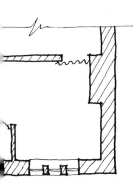

(Left) A typical finished drawing, the end result of the survey of a building.

PENNINE
FARMHOUSE.

Glossary

Bench mark: A permanent reference point of height created by the Ordnance Survey (illustration on page 26).

Datum point: The point of reference, chosen by a surveyor on the site, to which all measurements are related.

Dimensions: Measurements of length and breadth on land, of length, breadth, height and thickness in buildings and machines.

Dimensional survey: The measurement of surface features when surveying land, buildings and machinery.

Dumpy level: An instrument for measuring angles and determining levels.

Hand level: A cheaper version of a dumpy level because it is used without a tripod and lacks some other features.

Inclined measurement: A measurement taken at such an angle that it would be misleading if put on paper without correction.

Line band: Piece of string stretched tight to give a temporary base for measurements where no natural base exists.

Offsets: Measurements at right angles to a line band or natural base to allow distances to be measured at random points.

Plane table: A gadget which can be used to survey flat areas of land (illustrated on page 15).

Spirit level: A tool, which comes in all sizes, for ensuring that surfaces are absolutely level.

Spot level: The height of a particular spot, such as the doorstep of a house, in relation to other points on a site.

Ties: Additional measurements, other than the boundaries of a site or room, to serve as checks and to enable the final drawing to be made.

Introduction

'Local landmark to be demolished' is the kind of newspaper headline that prompts someone to make a survey. Britain has so many industrial structures, as well as all the non-industrial, that they cannot possibly all be surveyed by museum or other professional staff. So when the destruction of a site is announced, there is a sudden and urgent need to make a thorough record of what there is for future use. There is no virtue in trying to preserve everything that is old but it is not unreasonable to think that proper survey reports could be made. The cumulative total would add considerable detail to the picture of Britain's long industrial development.

Many people would like to make such a survey, and some excellent work is in print to show their success. As we are sure that many more people would try if only they knew what was needed and how to set about it we have set out to describe the basic methods of surveying, using short cuts wherever possible.

Do not be put off by the large number of methods. We have had to describe what to do for as many likely situations as we could include in a book of this length. Usually only one, sometimes two, methods will be right for a particular site. We suggest that you read through the book quickly the first time, to familiarise yourself with the kind of site each method is used for, and do not worry much about the method at this stage. Then read it again, trying to think of a local site where you would use each method, though you may not have examples for them all. Relate the method described to the site you know — where would you stand, which window would you measure from, what is the high point of the site, and so forth. What at first glance may seem a maze of methods will soon become acceptable common sense when you have fitted them to well-known surroundings.

Surveying mainly needs a methodical approach, patience, care and a stubborn streak that will make you turn field notes into a finished drawing. We offer the method — you can supply the rest. We are sure that you will find that it can also be fun, interesting and in the end satisfying. The original draft of this book was tried out with students and adult education groups; they enjoyed the work and produced some good results. We hope you will do both, too.

One final word, about metrication. Some of you will only be able to think in metric, others of you in imperial. Which should you use? For the sake of accuracy, use whichever you are most familiar with, as you are less likely to make mistakes. In ideal terms, there is much to be said for surveying buildings built since 1970 in metric and older ones in imperial, as each generation has made use of standard measurements in its day. It is very easy to miss the significance of some measurements by using different units: for example 5.0292 metres may not strike you as being anything much, but it is in fact the 5½ yards of the rod, pole or perch, which was the length of the ridge pole in cruck and ridge pole houses, and of the beam span in many cob buildings. We are very much of the opinion that, no matter what the survey was measured in, all drawings should be completed with a dual key in both metric and imperial measurements. This will make it easier to detect significant measurements of the kind described above. We have deliberately done some drawings in metric and others in imperial, so that you will feel at home whichever you best understand. Do not be deterred from undertaking a survey because you do not know your centimetres from your hectares (or barleycorns from furlongs) — what matters is having an accurate finished drawing with a key, so that anyone looking at it can work out from it what he wishes.

1. Surveying the land

To make a survey of any site you need to use a logical step-by-step approach to ensure that you record all the relevant information necessary to make a drawing at home. Surveys often have to be done as the bulldozers come over the hill. You cannot rely on being able to return to check on a dimension you missed the first time. Three kinds of land survey are needed but the equipment used is the same and so can be listed first:

1. A 15m (50ft) or 30m (100ft) measuring tape. (Metric-imperial equivalents given in this book are in most cases the standard size that equates with the other, not exact conversions.)
2. Four surveyor's ranging rods.
3. Four wooden pegs 75mm x 75mm x 450mm (3in x 3in x 18in), pointed at one end. A mallet or hammer.
4. String 30m (100ft) long, with a steel nut or some other weight tied on one end.
5. Sticks of chalk or yellow wax crayon.
6. A spirit level.
7. Adhesive tape.
8. A clipboard—a hardboard sheet 300mm x 450mm (12in x 18in), with a bulldog clip, paper and a sheet of polythene to keep off the rain.
9. A camera with black and white film.

Surveyor's chains measure very accurately but they are expensive and not suitable for surveying a building. For our purposes a measuring tape will be more suitable. These are made of steel, plastic or linen. Ranging rods are the red and white poles often seen on building sites. These can be expensive; a satisfactory substitute can be made from broom handles painted red and white at 300mm (1ft) intervals. Shape one end to a point to drive into the ground. Rods are mostly used for setting up a straight line by sighting three or more in a row across a site, though some other uses will be found for them. The wooden pegs are to mark more permanent points on the site. They are pointed on one end at an angle of 45 degrees on all four sides; by driving such a peg into the ground, an experienced surveyor can estimate the amount of loading the ground will take. The string will be used as a line band, for sounding wells and shafts, and to

make an accurate vertical line, among other things. The spirit level needs
to be as long as possible, ideally a builder's level a metre long. A useful
alternative can be a carpenter's level taped to a length of rigid board (e.g.
a length of floorboard), provided that both the long sides of the board are
parallel to each other. The use of a camera on site is discussed in Chapter
4.

Surveying an area of land falls into three distinct stages. These are:
measuring the site to show the surface distance of features from each
other (called the dimensional survey); working out the gradients of hills or
slopes (the levelling survey); and noting down any other relevant
information such as the kind of vegetation, soil and drainage
(topographical survey). We have written the book as if one person was
doing everything by himself, but clearly a group would divide the work
between them.

THE DIMENSIONAL SURVEY

The nature of the site will determine the choice of method to carry out
the survey, and sites can generally be classed as one of the following:

> (a) Small open site
> (b) Congested site
> (c) Small hilly site
> (d) Obstructed site
> (e) Large open site
> (f) Large hilly site

'Small' indicates a site where all dimensions are less than 30m (100ft).
Some site types could be measured by more than one method. For
example, (a) and (b) could both be measured with dimensions and ties,
while the plane-table method could be used for (a) as well as (e). Only one
method is given with each example to avoid repetition.

(a) Small open site

Start the survey by measuring the following dimensions (fig. 1): AB, BC,
CD, DE, EA, and the ties BE, BD, AC, CE. To plot the stream, first set up
the line XY with string, stretched tight. Measure YD, DX. Divide the line
into 3m (10ft) intervals. Make offsets from these points at right angles to
the line XY, and measure the distance of both banks of the stream from
this line. (See (d) for a quick way of setting up right angles.) Measure the
banks along ED and DC.

Before starting any drawing you must first choose a suitable scale. The

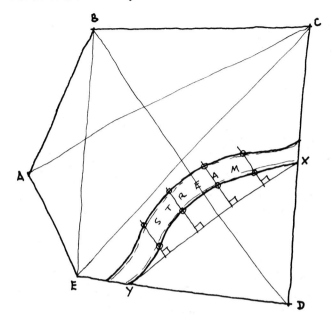

Fig. 1

usual imperial scales are 8, 16 or 32 feet to 1 inch, and in metric 1:100, 1:200 or 1:500. The largest scale (8ft to 1 inch or 1:100) will be best since all the dimensions are less than 100ft (30m).

Draw a base line horizontally on your paper and mark off the dimension ED to scale. Open a pair of compasses to the distance DC, place the point on D and make an arc at C, and then for the distance EC. Repeat the same procedure to mark in points A and B. Put in XY on your paper and scale it off into the intervals you had on the site. With a square mark in the offset lines at right angles to XY and so locate the positions of the banks of your stream. The solid lines can be inked in and the construction lines rubbed out.

(b) Congested site

This type of site is usually so heavily built up that dimensional ties are hard to obtain. A different method is therefore needed (fig. 2). Measure GH, HK, KL, LG. Also measure and mark off adjacent buildings forming the boundaries of the site. Measure the diagonal ties GB, AH, AG, BH, LF, KF, KC, LC, GF, CH and also AB, BC, CD, DE, EF and FA.

To draw the survey, start with the line AB as your base line and use compasses to find points H and G, as described in method (a). Proceed from there to locate the points L and K. Now check point H by opening your compasses to the scale distance GH, placing the point on G and making an arc at H, and double-check by repeating with the dimension KH. This is why cross ties are necessary. Always try on the site to tie each point in with two or preferably three measurements so that you can check points as you draw them without having to return to the site to remeasure.

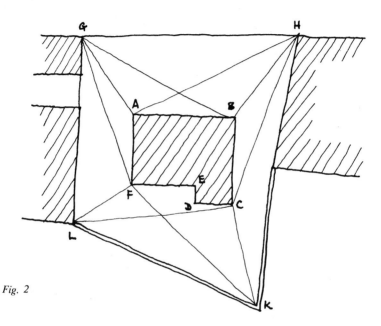

Fig. 2

(c) Small hilly site

The hill (fig. 3) makes it impossible to take direct cross ties in this case, so arbitrary points have to be created, as at X. Ranging rods or pegs are driven into the ground in places which give a clear view of at least two other points; cross ties can then be measured normally. Measure AB, BC, CD, DE, EA, EX, DX, and the diagonals AC, CX, XA, which tie in the other points. To find the position of the highest point of the hill, fix a ranging rod at point Y and measure BY, CY, DY, and EY.

As fig. 4 shows, there is a difference between an inclined measurement and the real horizontal measurement. Trigonometry tables will be needed to make the necessary corrections; they can be obtained through most libraries or bookshops. Measure the angle Z, and use cosine or sine tables to work out one of the following equations:

$$\text{horizontal measurement} = \frac{\text{inclined measurement}}{\text{cosine Z}}$$

or

$$\text{horizontal measurement} = \frac{\text{inclined measurement}}{\text{sine (90 - Z)}}$$

Methods of measuring the angle Z are described on pages 19 and 22. Drawing this survey raises no new problems.

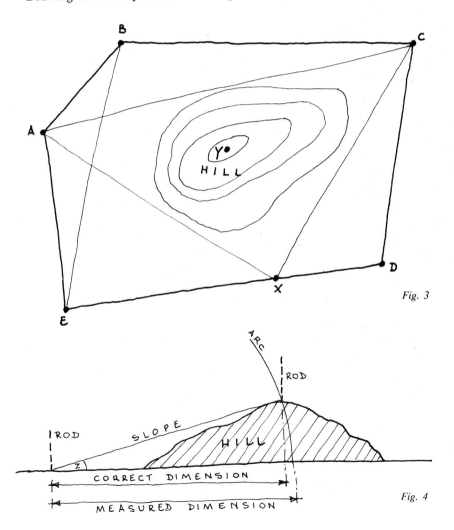

Fig. 3

Fig. 4

(d) Obstructed site

This situation arises on a site where there is an obstruction that prevents a dimension being taken between two vital points (fig. 5).

Where a building obstructs the measuring of the line AB you cannot rely on adding AD, BC and the gable end of the building in case the walls of the building are not parallel. This would inevitably mean that the gable was different from the distance between the walls on the line AB, and you would have a false measurement. Instead use four ranging rods to set the line EF at right angles to BC, and HG at right angles to AD. EF must be the same length as HG. Measure BK, EH, and AL, and add the three dimensions together to give AB. The building can be located by taking dimensional ties from E, H, K and L to the corners of the building, as previously described for congested sites.

To set out a right angle simply, you need a measuring tape and either two people or two ranging rods — people will be quicker (fig. 6). Hold the tape at zero; the second man holds it at the 3m (9ft) mark, the third man at 8m (24ft) and you again at 12m (36ft). A right angle is formed where you stand when the tape is pulled tight.

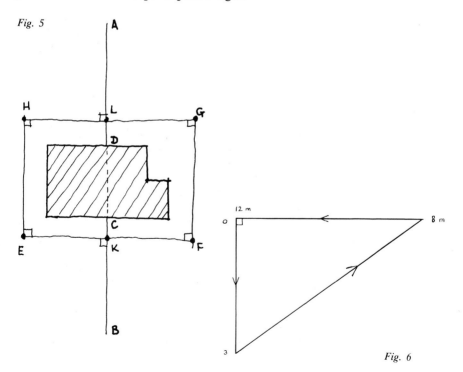

Fig. 5

Fig. 6

Fig. 7

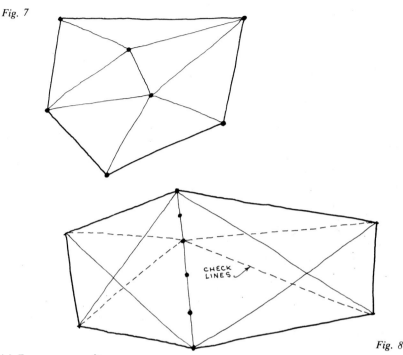

Fig. 8

(e) Large open site

It is not possible to take full dimensions with the 30m (100ft) tape on large sites. You could try tying two tapes together with a piece of string, but this makes difficulties — it is more awkward to get the tape in a straight line on the ground, and calculating the measurements offers every chance of making mistakes. A more desirable way of handling this kind of site is to divide it into smaller sites by setting up arbitrary points with the ranging rods so that no dimension is over 30m (100ft). There are two ways of doing this, as shown in figs. 7 and 8. Place two or more rods in the field so that no diagonal is more than 30m (100ft), then measure as for small open sites (fig. 7). There is no virtue in overdoing the planting of rods; the site should still be uncluttered, not a garish forest.

An alternative method is to divide the site into two (or more) with a straight line of ranging rods and to treat the survey as two small sites (fig. 8). Care must be taken to make check ties back to the rods in both parts of the site because the rods are the only feature that links the parts together. The line of rods is the base line when you start to make the final drawing.

Large and reasonably flat sites can also be surveyed quickly by doing a plane-table survey. A plane table consists of a sighting arm on a central pivot on a board, which is in turn mounted on a tripod or firm stake (fig. 9). The board must stay in the same position throughout the survey. A piece of paper is fastened to the board with clips or adhesive tape so that it cannot move, and the sighting arm is replaced in its pivot hole. To make a survey, first set the table up in the middle of the site, making sure that every feature to be included can be seen. Features that are indistinct, such as bends in boundary walls, can be made more obvious by ranging rods. Make a rough sketch of the site. Line the arm up with the first point to be noted by sighting through the arm to the feature. Make a mark on the paper beside the pointer on the arm, number or letter the point and write the same symbol by the feature on the rough sketch. Continue round the site by turning the arm from one feature to the next. Then take the dimensions from the centre pivot to each feature that has been recorded on the plane table. Also measure the distances from point to point around the boundary — these will serve as checks when you make the final drawing.

To make the drawing, fasten the survey paper from the plane table to your drawing paper with adhesive tape or drawing pins. Project lines from the pivot through the pencil marks, and scale off along these the dimensions you measured on site. This will give you the shape of the site to scale, which can be checked by using the boundary dimensions (fig. 10). The site could be split in two if it is large and two plane-table surveys made, so long as each of them includes the row of ranging rods.

(f) Large hilly site

This is usually the most difficult type of site to measure, and the only practicable way is to use a theodolite or surveyor's level. You may know an architect or surveyor who would lend you the equipment and you might persuade him to help as well. Using wooden pegs, set up the points A, B, C, D and E a metre inside the boundary wall so as to allow room to set up the equipment within the site (fig. 11). Fix a ranging rod in the summit of the hill. With the theodolite measure the horizontal angles ABC, BCD, CDE, DEA, EAB on the protractor scale of the instrument. Similarly, calculate the horizontal angles CDF, etc. When making the drawing, use a protractor and scale ruler to plot in the dimensions and angles known, remembering that ABCDE is a metre inside the real boundary of the site. This will give you a scale drawing of the site, and it will be straightforward to mark in the summit of the hill. A plane-table survey is not suitable for this kind of site as all the dimensions to the centre pivot will be sloping, probably at different angles. The dimensions will require varying amounts of correction and there is little chance of double-checking if errors creep in.

If only a small part of a hilly site is needed in the survey it can be staked

Fig. 9. Plane table

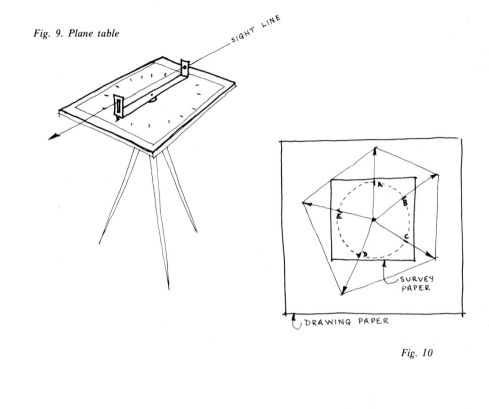

Fig. 10

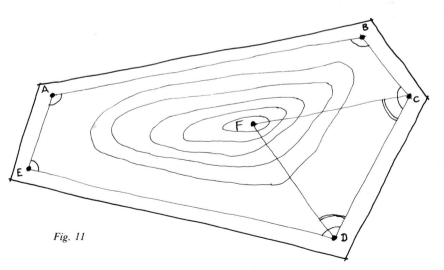

Fig. 11

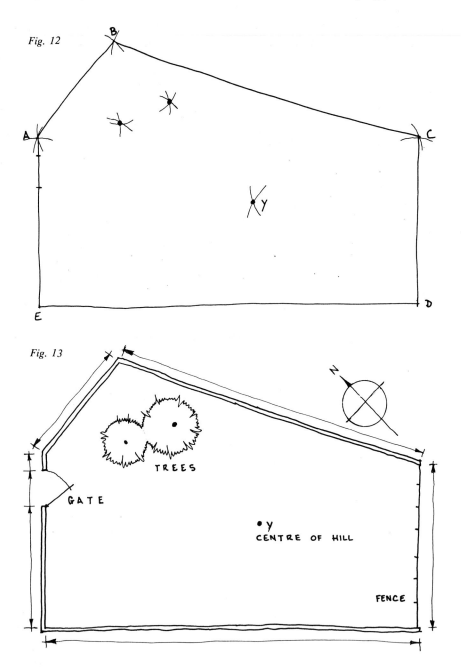

Fig. 12

Fig. 13

off and treated as a small hilly site. This part will be surveyed in detail (see page 10), while the remainder of the site can be built up from Ordnance Survey maps. It will be necessary to prepare for this by taking in some features during the survey which are clearly marked on the OS sheet. Remember to check whether the OS map is metric or imperial.

Completing the drawing for all kinds of site

Using the compasses, scale rulers, protractors and so forth will have given an outline of each of the kinds of site described in this chapter, and the drawing at this point will resemble fig. 12. It is no more than a geometrical shape and lacks detail. First remove the compass arcs and other construction lines. Add to the bare outline the walls, fences, hedges and other features of the site, and write in the dimensions shown and a north point. Later a levelling survey and topographical survey will be added to accompany this drawing, and the growing collection will gradually begin to present a fuller picture of what the site was really like (fig. 13).

You may wish to calculate the area of the site when you have completed the drawing. A method for irregular shapes is to divide the shape into parallel strips of equal width (fig. 14) and calculate the following sum:

$$\frac{AB + YZ}{2} + CD + EF + GH \text{ etc to WX} \times \frac{L}{N}$$

where L is the distance between AB and YZ and N is the number of strips. Sites with straight boundaries can be dealt with by dividing them into squares, rectangles and right-angled triangles, calculating each separately and adding them together.

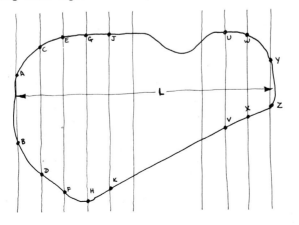

Fig. 14

Check list

There is nothing more annoying than starting to make the final drawing at home only to find that you have omitted a vital measurement. You cannot guess at it, and the only alternatives are a return to the site or an inaccurate drawing, quite apart from the waste of time. The remedy is to work out a step-by-step approach to making a survey and to use the same method every time. The following list is an outline guide which you may wish to lengthen, or you may prefer to compile your own. Our list is not the only way it can be done but some such methodical approach is essential to avoid irritating gaps in your field notes and delays in completing the drawings.

1. Sketch shape of site and note any walls, trees, streams, holes, buildings, etc. Mark on it the north point.
2. Make up your mind which kind of site it is and choose the appropriate survey techniques.
3. Measure boundaries and set up ranging rods required for lines or offsets.
4. Measure cross ties and diagonal ties which will be needed for double checking. Make a note of any seriously sloping dimensions so that you can correct them after the levelling survey.
5. Measure the outsides of buildings and their distance from site boundaries.
6. Mark in trees, telegraph poles, etc. (Measure to centre of trunk.)
7. Measure gateways and note which way they open.
8. Note thickness of boundaries and their construction — stone, brick, hedge, etc.
9. Measure width of roads and verges adjacent to site.
10. Note position of any manholes, hydrants, etc.
11. Locate any buildings adjacent to site boundaries.
12. When making a plane-table survey, be sure to record what each mark stands for.
13. If two tapes are tied together, double-check that you have calculated the tape readings correctly.
14. Make a note of the position of any trial holes you have made. (See page 29).

THE LEVELLING SURVEY

The object of this survey is to work out the gradients of the site. The end result will be a drawing of the site which shows contours and the heights of particular points or features (fig. 15). A small site can be surveyed with a spirit level and a long board, and the result can be quite accurate if the survey is done with care and patience (figs. 16 and 17). Sharing the work among several people would be the best arrangement. Finding the levels of a larger site, or any site where accuracy within 20 millimetres or less is needed, calls for the use of more expensive equipment such as a hand level or a dumpy level. Both are optical instruments. A dumpy level is mounted on a tripod, and the person using it takes readings from a calibrated ranging rod or staff held by someone else (fig. 18).

A hand level, as may be guessed, is really a hand-held version and somewhat simpler (figs. 19 and 20). Both the staff and levels are expensive. If you need that kind of accuracy, it would seem sensible to enlist the help of a professional surveyor, rather than to spend money on costly equipment that needs some training to use accurately. The author once led a party of students to survey a cluster of houses which were on the

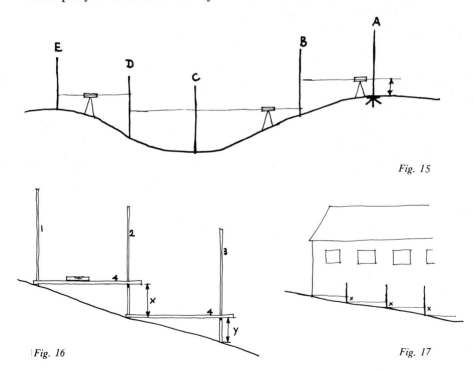

Fig. 15

Fig. 16

Fig. 17

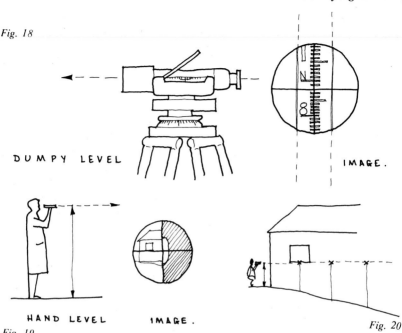

Fig. 18

DUMPY LEVEL

IMAGE.

HAND LEVEL   IMAGE.

Fig. 19

Fig. 20

point of being demolished. They were grouped around a hillside spring, and no gradient remained constant for many yards. The survey had to be done at speed and there was only time to survey the buildings, yet a plan of their arrangement would appear very haphazard without reference to a levelling survey. On one visit we found the council surveyor making the survey for redevelopment purposes and the council were only too willing to make a photocopy of the survey available to us when it was completed. Most sites that are being cleared for redevelopment will also be surveyed, and most developers are prepared to listen to requests that arise from genuine interest. Nothing much will be gained from duplicating the work of professionals and it is easier to mark in the few points that were of no interest to the developer than to do the whole survey yourself.

Not all sites, however, are threatened by redevelopment, and for those that are not you must make your own survey. There are two main methods. One is based on spot levels taken at random points and the other on levels taken at points determined on a grid system laid out across the site. Spot levels at particular points are often all that is needed to show, for example, the relative heights of a building's corners at ground level, and these in relation to the high, access and low points of the site. A grid survey is needed if the contours have to be calculated. Having decided which kind of survey is needed for the particular site you are at, first draw a rough cross-section of the line of the survey so that you can relate your figures to points along the line. This saves hours of puzzling over lists of figures sometime later when you are trying to convert them into a drawing, and it is often a help when making the actual survey.

Spot levels

There are two ways of finding the spot levels A, B, C, D, E, F, G, H, K, L in fig. 21. One is to use a hand level or something similar. Start from the lowest point of the site, say G. Sight on to H, B, A, F and L, making use of an assistant to hold the ranging rod in the site corners and to make chalk marks on walls (fig. 22). Make a note of the distance to the ground at each point, and mark the measurements on the sketch of the cross-section. Then go to L in order to take the levels at the other side of the building. Draw another cross-section of F, E, D, C, K, and sight on to F; called the backsight, this is essential to be able to relate the two cross-sections

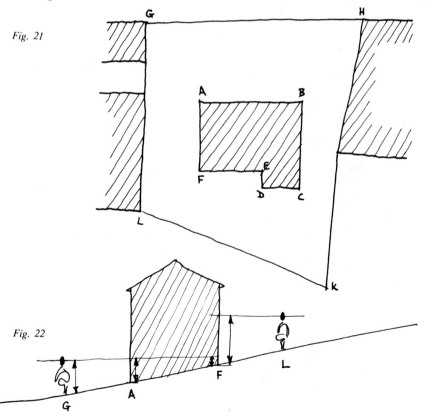

Fig. 21

Fig. 22

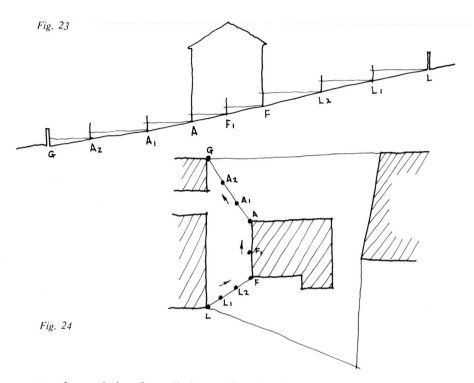

Fig. 23

Fig. 24

together and therefore all the readings back to the starting point at G. Now make the readings (foresights) to E, D, C and K. The sketches of the cross-sections are of considerable use as an explanation of how to interpret your results.

This kind of site can also be surveyed with a spirit level, and the result will be just as accurate if it is done carefully. No amount of expensive equipment will remove the need for care and patience. Start at the top of the site, as it is easier to work downhill. You will need to change points regularly unless you have a very long and rigid plank and some strong assistants. The changes must be on a straight line from one spot-level point to another—it may be useful to stretch a line between pegs to guide you (fig. 23). Starting from the highest point (L), set the plank level and make a note of the length from L to the first change point L1 and the vertical distance from the ground to the underside of the level (or, of course, of the board) (fig. 24). Continue for the whole site, making sure that all levels are related back to L. Only the spot levels at L, F, A, etc, will need to be calculated and marked in when the final drawing is made; L1, L2, etc, can be ignored.

Grids and contouring

In an open field there are few spot levels close to each other which can be easily identified, so points must be artificially constructed. They are set out at regular intervals, which makes them easy to locate on a drawing. The use of a line of ranging rods and the method of tying them in so that they could be located later were described on page 10. Alternatively, mark 10m or 5m (30ft or 15ft) intervals along one boundary wall, which will make just as good a starting line provided the boundary is straight. The intervals will need to be less if the land is steep. Once you have your straight line of ranging rods, take spirit levels (or instrument levels) from A to F (fig. 25). Set up a line band GM parallel to AF and the same distance from it as the intervals between the ranging rods — each grid must be square. Mark H at right angles to B, and so on along the string. Move the ranging rods to new positions and again take spirit levels from G to M. Also take a level from A to G. Continue in this way until the whole site has been covered. Always take a level from each new line of rods back to A, which is the starting point of the survey. Not only does this tie each new series of levels to an identifiable point (in this case the angle of the boundary), but by going back to A each time you greatly reduce the chance of compounding errors, whereas, if you had made a mistake in levelling G, any levels later related back to G would also be wrong.

You will end up with a series of points which can be located and related to your dimensional drawing. The levels at each of these artificial points are not of much use as they stand. It is possible to use them to draw to scale an accurate section of the site along any of the grid lines. Better still, they can be converted into contours, which are more helpful. The method

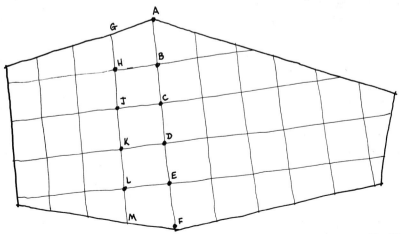

Fig. 25

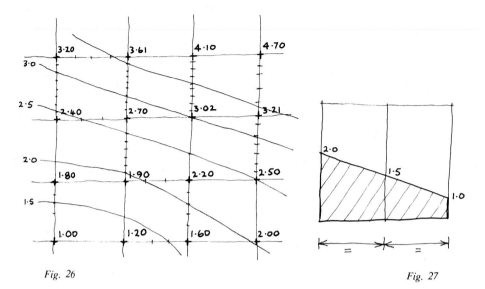

Fig. 26

Fig. 27

is to decide on the distance between contours, say 1m, and join all points of equal height. This presents problems of estimation or interpolation, which are soon overcome with a little practice (see figs. 26 and 27). All the heights are related to the lowest point of the site. This is perhaps the more straightforward way to proceed. It is just as possible to start from the highest point on the site, though the figures will be reversed (fig. 28). The final results will be the same but avoid using both practices to make the same drawing unless you can keep a cool head. You can sometimes save yourself some work at the contour-estimating stage if the nature of the site allows you to combine the starting point of the survey with the lowest

Fig. 28

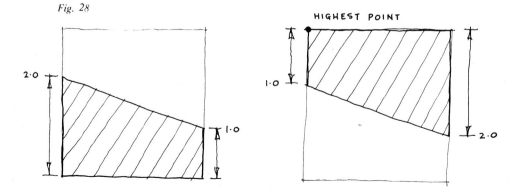

HIGHEST POINT

point. If this is taken as 0m, all verticals on the first line of rods will be more than nought. The first point on the second row will also be more than nought, say 0.5m; if each point in the row is added to (or subtracted from) this and the process continued throughout the site, you will have no arithmetic left to do to relate the various parts of the grid together. The same method can be used from the highest point, in which case all measurements will be subtractions from the first figure (except on bumpy sites), which will have to be an arbitrary figure large enough to allow for all the differences on the site. Alternatively start from nought and treat all measurements as steps down from nought. It matters little which you do as long as you use only one method at a time. On some sites only the land immediately round a building is of interest, in which case you need only set up a grid around it.

Datum points and bench marks

Every survey must have a datum point in order to relate all levels to one point. In the methods described so far every level has been related to the starting point. This means that, where the starting point was a ranging rod, the point cannot be found later when the rod is removed, so that the levels are related to nothing.

It is therefore necessary to create a temporary bench mark. You can make this point an object which cannot or is not likely to be moved, such as a gatepost base, or make a permanent mark on a wall, or use a doorstep. Level back from your starting point to this position, and the survey is at once related to a permanent point on the site. This temporary bench mark is usually called 100 (fig. 29), and all other levels are adjusted to relate to this point. The survey drawing should look like fig. 30 when all the figures have been added.

At some future date, it may become necessary to relate the site levels to Ordnance Survey levels. This can be done by locating an OS bench mark. The location of bench marks can be found on 6in (1:10,000) or 25in

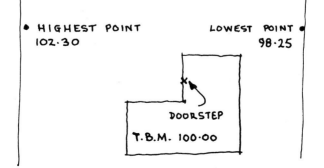

Fig. 29

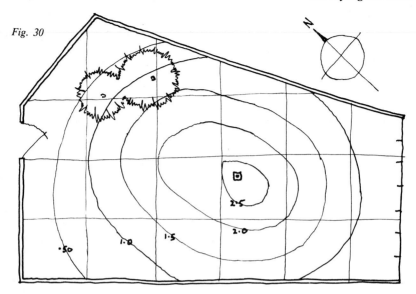

Fig. 30

(1:2,500) OS maps; they are marked on walls or other fixed points as in fig. 31. They are related back to sea level at Liverpool or Newlyn — a note at the foot of the map will say which. The older Liverpool level is not as reliable as the Newlyn level, but both serve the present purpose equally well. Having found a bench mark, probably outside the site, take levels from it to your temporary bench mark (this highlights the value of making it as permanent a feature as possible). After that, adjust the figures arithmetically to find the proper levels. The site contours may need to be revised unless they were done to the same intervals as those used on OS maps. But beware — the Ordnance Survey is replacing the maps in imperial measure with metric and both editions will be in use for some time yet.

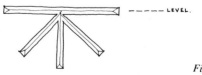

Fig. 31

Check list
1. Decide on the type of survey to be used and proceed accordingly.
2. Make a sketch of the positions of spot points or grid, and clearly mark each point, the starting point and temporary bench mark.
3. Check that all points have been related back to starting point and starting point back to temporary bench mark.

4. Make separate sketches of each section of levels to explain exactly what you did, and relate each sketch to another.
5. Note position of bench mark (on map) and its level for future reference.
6. Measure depth of wells or pits by string and weight. Take level at top of well (ground level) and relate depth to this.
7. Measure depth of stream and levels of banks.
8. Take levels of tops of manhole covers and measure depth to bottom of drain.
9. Find ground floor level of any building by taking level on top surface of doorstep.

THE TOPOGRAPHICAL SURVEY

The final stage in a land survey is to make a site analysis diagram. You will need either a large scale OS map, a print of the measured survey you have made or a sketch of the site. Take it to the site and mark on it from observation the important features suggested by the check list below, using notes or the kind of symbols given in fig. 32. This survey may help to show why the site developed in the particular way that it did, and the factors that most affected its development. It also takes in some kinds of information that would be omitted from dimensional and levelling surveys.

Check list
1. Show all trees and shrubs, and the names of the trees if possible. Some varieties were planted to dry out marshy land. Trees also protect from sun, wind and traffic noise and give privacy or obstruct the view: note which.
2. Which are the best views and the obscured views from the site? Have they any bearing on the layout of the site?
3. Check and mark any evidence of flooding and swampy ground.
4. Check for nuisance factors outside the site which might affect it, e.g. outfall from a power station or dye works.
5. Points of pedestrian and vehicular access to the site.
6. Hills and hollows which may affect protection of site from weather, cause frost pockets or give rise to mist.
7. Diagram of prevailing winds, travel of sun and north point. Avoid the fashionable but confusing symbol in which the N is placed at the opposite end to the north-pointing arrow (see right).
8. Springs, wells and watercourses.
9. Ground cover — vegetation, eroded areas, subsidence.
10. Type of subsoil and topsoil (see note on trial holes below).
11. Positions from which photographs of the site are taken.

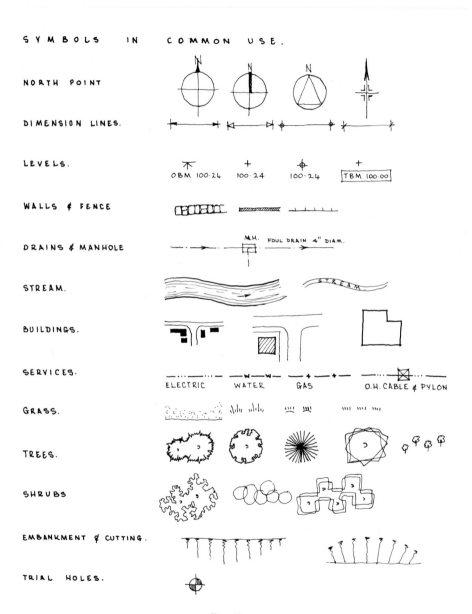

SYMBOLS   IN   COMMON USE.

NORTH POINT

DIMENSION LINES.

LEVELS.

OBM 100·24   100·24   100·24   TBM 100·00

WALLS & FENCE

DRAINS & MANHOLE

M.H.   FOUL DRAIN 4" DIAM.

STREAM.

STREAM.

BUILDINGS.

SERVICES.

ELECTRIC   WATER   GAS   O.H. CABLE & PYLON

GRASS.

TREES.

SHRUBS

EMBANKMENT & CUTTING.

TRIAL HOLES.

Fig. 32

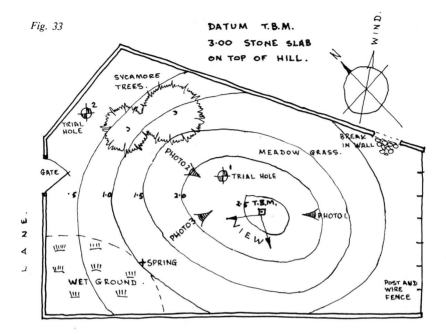

Fig. 33

DATUM T.B.M.
3·00 STONE SLAB
ON TOP OF HILL.

Some of these items have been mentioned before but can sometimes be overlooked. Ask the owner of the site before starting to dig any trial holes. These should be a metre square and deep. The strata revealed in the sides will generally be sufficient to tell what the foundations of an adjacent building are bearing on, the reason why particular kinds of vegetation are common, and so forth. Notice whether any such holes fill with water and, if so, how fast they fill and to what height. This will give further information about the subsoil and the height of the water table. Check the information obtained from trial holes with a geological drift map for the area, which shows where the various strata drift to the surface. Your final land drawing should now look like fig. 33.

Additional information for the file and where to find it:
1. Position of gas mains — gas board.
   Position of electricity mains — electricity board.
   Position of GPO lines — Post Office.
   Position of water mains — water authority.
   Position of main sewers — local authority, main drainage department.
2. Future developments of mains — as above.
3. Future planning development — local authority.

 4. Easements and rights of way — OS maps, deed plans.
 5. Zoning of land, building restrictions — local authority.
 6. Mineral rights, and records of previous mining activity — local mineral valuation officer and National Coal Board.
 7. Geology of site — OS maps, public library.
 8. Water analysis — public analyst.
 9. Weather records for the area — meteorological office or museum.
10. Previous use of site: basement walls, wells, shafts, quarries — old local authority maps.
11. Adjoining owners' names and addresses.
12. List of photographs, with apertures, name of subject and site reference number.

Local authorities, gas boards, etc, are usually willing to help with information of this kind for genuine recording work. It helps if you send a copy of the site plan to the office for that area so that they can mark on it the information you request. This is additional work for them, so be patient — 'please' and 'thank you' work wonders.

2. Surveying a building

One of the most common needs for a quick survey is news that a building is to be demolished or, perhaps, modernised. In these circumstances it is important that all the measurements are taken that will make it possible to draw plans and elevations of the destroyed building. These remarks apply just as much to details as to main dimensions — it is fatal to try to memorise that there was a light switch 60mm from the door in one room, a gas fire in another and a sink slightly off-centre under the window in a third. Equally, it is optimistic to hope that you can return to the building to check these details when making the final drawing. By the time most drawings are reaching the stage of having the details added, the building is down and there are curtains at the windows of the new block of flats on the site. It takes very few minutes extra to mark on details when making a survey, and the information may be important later. If in doubt, measure it and make a note. It is always better to have too much information than not enough.

In writing this chapter on surveying buildings we have assumed that you have read and understood the previous chapter, so as to avoid repetition. Two points, however, can never be stressed too often. All buildings, even empty ones, belong to someone, and you should make every effort to find the owner and obtain permission to enter the building before you carry out a survey. Most owners are pleased to help, though you will sometimes have to accept a refusal. It will help if in your approach to the owner you do not assume that you have a right to carry out a survey on his property. If you choose to proceed without permission, you will get nowhere by trying to plead that the survey was in the national interest and you run the risk of being charged with trespass.

The second point concerns safety. Empty buildings are seldom left in a sound condition. Floorboards are likely to be rotten, roof slates loose, door and window frames liable to fall out, and so forth. For this reason, as well as for other help, it is always better to take someone with you when making a survey — two heads can warn each other.

Equipment required
15m or 30m tape (50ft or 100ft).
A 2m (6ft) spring steel rule, or a folding rule which can be held in an upright position.
A ball of string and a plumb bob (or nut and bolt).

Sticks of chalk or yellow crayon.

A spirit level.

A piece of hardboard 300mm x 450mm (12in x 18in), bulldog clip and
sheet of polythene the same size to cover drawings when working out of
doors in the wet.

Paper, pencils and ruler.

A torch.

Camera, with black and white film.

A strip of lead 25mm (1in) wide and 450mm (18in) long.

Buildings can be grouped into four categories for survey purposes.

(a) Average-sized buildings capable of being measured easily both inside
and out, such as houses, farm buildings and small factories.

(b) The ruins of such buildings, which can only be measured outside.

(c) Tall buildings only capable of being measured internally, such as
mills and warehouses.

(d) Tall structures which can only be measured externally, such as
chimneys, obelisks and bridges.

Each category calls for slightly different techniques. Most buildings to be
surveyed, however, fit into group (a), so surveying a building of this kind
will be explained in detail. Later sections will cover the other categories
only in so far as they differ from this group.

(a) Average-sized buildings

Start the survey by making a sketch of the interior of the building. This
must be large enough to allow you to mark on it all the measurements you
will make. The sketch could thus be either a complete floor plan or a series
of sketches for the separate rooms on one floor (fig. 34). If you use several
sketches for one floor, you will have to relate them to one another on a
further sketch either by a single measurement through the open doors on
that floor or by a continuous measurement along the outside of the
building. Mark in on the rough plan all window openings and mullions,
the door openings and the directions the doors swing. Chimney breasts,
steps and sinks should also be marked — anything, in fact, that is a
permanent feature of the room or floor. Number each tread of the steps,
not forgetting that the landing they lead to is also the top step. Show the
direction of the flight of steps out of the room with an arrow and *up* or *dn*.
Add to the rough sketch the line of the external walls. Fig. 34 indicates
how the sketch should look at this stage. The more careful you are to draw
the proportions accurately, the easier your work will be in later stages.

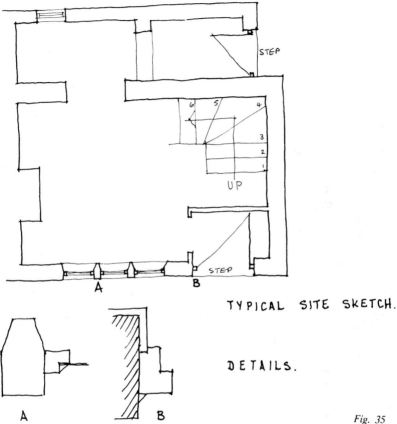

Fig. 34

TYPICAL SITE SKETCH.

DETAILS.

Fig. 35

Where you need to know the shapes of mouldings around doorways, for example, or the position of window frames between mullions, it will be necessary to make large-scale sketches to allow the marking in of many fine measurements (fig. 35). Relate these large-scale drawings to the main sketch (by a series of letters, for example) if there are many of them.

Start measuring the floor or room in your sketch in an orderly way, so that you do not omit any dimension. This is a habit that must become second nature, as any other method inevitably leads to missed measurements and disappointing results. It is often best to start measuring from the point at which you enter the room, go round clockwise and finish where you started. Make all your dimensions as accurate as possible — to the nearest 5mm (quarter inch). Each measurement will

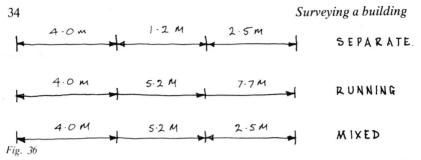

Fig. 36

have to be separate if you are working alone, which cannot be helped. If there are two or three of you, however, you can use a running dimension and this considerably reduces the opportunities for making mistakes (fig. 36). It is also much quicker.

To make a running dimension along one wall, two people will need to hold the tape tight against each end wall. It must be parallel to the wall being measured, level and free from sag, all of which would distort the measurements. (It is sometimes possible to hook the tape behind a board at one end, freeing a person for another task, but be careful to note whether the tape is shortened or lengthened by this expedient.) The measurements are then read off at each mullion, door frame or whatever, each figure being higher than the last. It is most important that you mark the arrows correctly, as in fig. 36. Should you ever have to mix separate dimensions and running dimensions on the same sketch (and it is better not to), be extra careful how you mark the arrows. Mixing dimensions should rarely be necessary; separate dimensions can be worked out very simply from a running dimension if they are ever needed, and that is better than risking confusion by using different systems on the same sketch. The widths of windows and doors should be taken as the full opening made by the mason. If door frames are masked with heavy mouldings it will be necessary to estimate where the opening ends (fig. 37).

Complete all the horizontal measurements round the room. The next stage is to take cross ties. Start with the diagonals from opposite corners. Then take extra ties to the window jambs, door frame, chimney breast or other permanent features (fig. 38). These will serve both to check the accuracy of your earlier measurements and also to show up any unusual angles or shapes which were not apparent otherwise.

The next dimensions required are heights. Measure from floor to

TO ESTIMATE MASONS OPENING AT DOORWAYS.

Fig. 37

Fig. 38

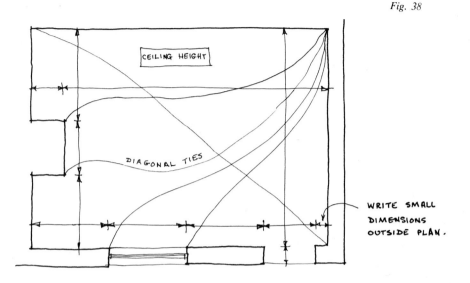

TYPICAL SKETCH OF DIMENSION LINES.

ceiling, and also the heights of window sills, window heads and door heads. For the stairs, measure the risers (avoiding the worn centres of the treads) and add them together to give the vertical height of the staircase. As a check to this, or to measure uneven steps, it is best to measure a vertical tie (fig. 39). To do this, measure the height from the floor to any convenient point on the staircase (fig. 40). Draw an accurate horizontal line from this point, using a spirit level. Now measure from this line to the ceiling above. Adding the two together will give the dimension AC. Subtracting DE from AC will give the vertical rise of the stairs. As FG has already been measured, subtracting DE and FG from AC will give the thickness of the floor as further information.

The final measurement is the thickness of internal walls between rooms. This may be possible in a doorway but if there are extensive mouldings around the doorway and where there is no door in the wall an estimate will have to be made. If details of light switches and the like are needed they are best inserted on a new sketch of the room, using the symbols given at the end of this chapter, unless you wish to devise your own.

The same procedure should be used for each room and floor of the building. It is important to relate the measurements of rooms to each

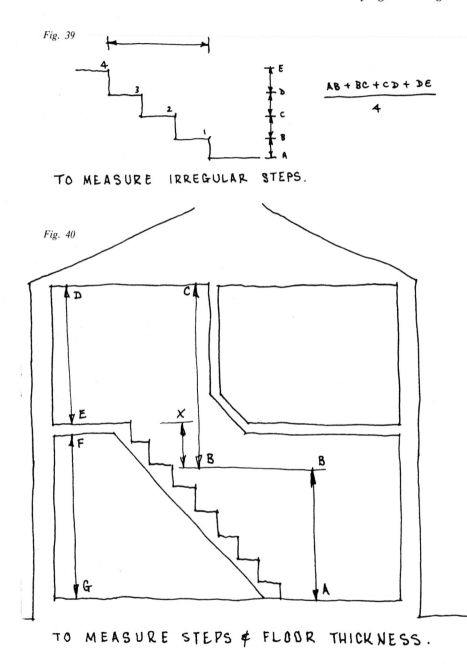

Fig. 39

$$\frac{AB + BC + CD + DE}{4}$$

TO MEASURE IRREGULAR STEPS.

Fig. 40

TO MEASURE STEPS & FLOOR THICKNESS.

other, particularly if there is a sketch for each room. In such a case, it would be wise to have another sketch of the whole floor of the building, with a running dimension through all the rooms (fig. 41). There is far less chance of errors following this precaution as there is only one real dimension and only once chance of an error. If separate dimensions are used, errors can creep in at B, C and D, increased by the number of rooms on this floor. (The running dimension may also provide the thickness of internal walls in some cases.)

Structural details should also be noted, using quick additional sketches as necessary. Determine the size of the floorboards and joists and the spacing of the joists, which will be indicated by the row of nailheads in the floorboards. Note also the spacing of ceiling joists where these are not part of the floor above. There is normally a manhole giving access to the roofspace. This allows you to sketch the roof trusses and other details of the construction of the roof. For example, are the slates held on by nails or wooden pegs, what distances separate the laths which support them, are the trusses king post, queen post or what, and so forth? It should be possible to measure at least the lower parts of the trusses, and also the distance between trusses and between truss and end wall. Be careful only to stand on joists — ceiling plaster was never meant as a floor.

Take the dimensions of any concrete or steel beams, trying to imagine which are load-bearing and which are not (fig. 42). This will lead to an appreciation of the structural principals of how the loads are transmitted to the ground in this particular building, which may in turn lead you to notice features which you might otherwise have overlooked.

You will need to know the thickness of the external walls when you come to make the final drawing. If possible, measure at a ground-floor

Fig. 41

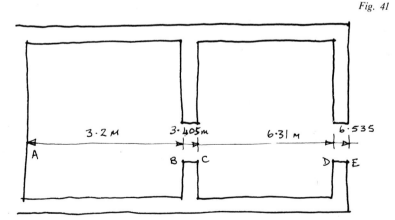

Fig. 42

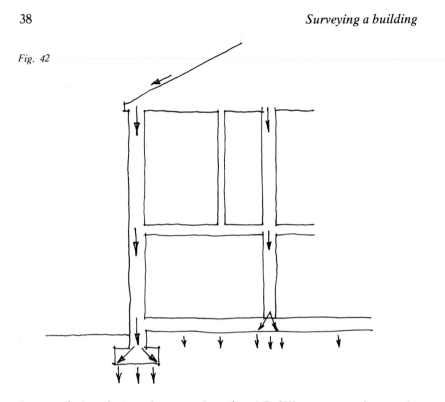

door or window that can be opened, noting AD. Where no opening can be
found, measure AB and CD, estimate the thickness of the frame and add
the three dimensions to give AD (fig. 43). It is worth making a second
measurement at first-floor level, as some walls are thicker near the ground
where they carry more weight. Measure steps up to the door, so that you
know how much the floor level is above ground level. That completes the
survey inside the building.

Start the exterior survey by making rough sketches of the elevations,

Fig. 43

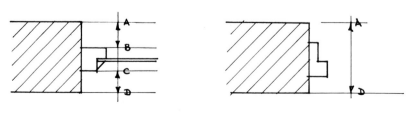

WALL THICKNESS AT WINDOWS & DOORS.

Fig. 44

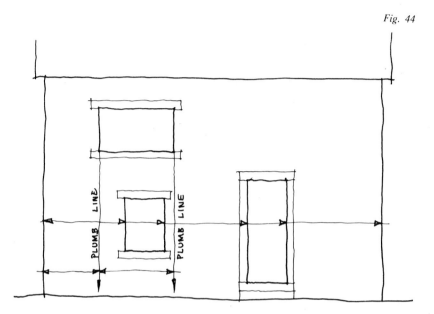

showing only the basic necessary details. Start measuring from one corner and go round the building in an anticlockwise direction. Use a running dimension as shown in the diagram, taking in all door and window openings. Use a plumb bob from the jamb of each window where possible to obtain accurate measurements of upper-floor windows (fig. 44). Where this cannot be done, estimate their position with chalk marks lower down and measure from those.

It is possible to calculate vertical dimensions from the internal survey but greater accuracy will be obtained if a separate measurement can be made. The easiest way to measure the height is from a ladder, if you can borrow one nearby. Where this is not possible, open an upstairs window and use the spring rule to measure A. Hold the tape at the head of the window so that someone outside can read the dimensions B, C, D and E (fig. 45). It may be that this cannot be done either. If this is so, only a visual estimate can be made externally but even this may help to check a dimension calculated from the inside measurements. Some standard building products can help in making a visual estimate. In brickwork, for example, measure four bricks and four joints to see if they come to twelve or thirteen inches, or to a local gauge. Count the number of courses and work it out from there. Rainwater pipes are usually in 6ft lengths (1.83m), which can be counted. Domestic stone window sills, heads and jambs are

Fig. 45

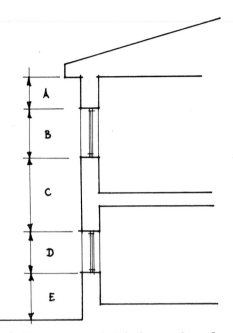

usually 6in thick (150mm). Note on your sketch the number of courses of roof tiles or slates.

Having done the best you can, photograph each elevation in case you have missed a detail. Use black and white — a pretty picture on a screen is no help when what you need is a photograph beside you to make the final drawing. Photographs of the gable end, together with any details you were able to note while in the roofspace, should make it possible to calculate the height of the roof from the eaves to the ridge.

You may be faced with decorative stone mouldings that you would like to record. If you are unable to sketch them accurately freehand, then carefully press the strip of lead on to the moulding, and just as carefully pull it away. Place it edge-down on to a piece of paper and run a pencil down the side of the lead that touched the moulding (fig. 46). Photograph the moulding also, to show its relationship with the rest of the building and to guard against having missed some details connected with it. A word of warning about mouldings — these are frequently protected with a lead cover. It is quite possible for the stone mouldings to be cracked under the lead or to have fallen away and yet leave the lead in its original shape. Therefore beware of standing on any lead coverings.

That completes the survey for the great majority of buildings in this group, so you can go home with a sense of something accomplished. Some

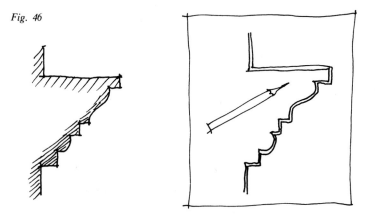

Fig. 46

USE OF LEAD STRIP ON MOULDINGS.

buildings, though, have curved walls, others have arches, and the technique for surveying them is the same (fig. 47). For internal walls and arches, measure AD and divide it into equal parts, AB and BD. Measure BC. If it equals AB, the recess or arch is semicircular, and no further measurement is required. In other cases, where the recess is a segment of a circle, measure AD and divide it into AB and BD as above. Draw the line BC at right angles to AD (using chalk marks on the floor) and measure AC and CD. (When you come to draw this again, set up the dimensions as sketched. Bisect AC and CD into equal parts and draw lines at right angles to point E. This is the centre of the circle and you can use a compass to draw the arc between A and D.)

Curves on external walls and the outside of arches can be measured by constructing straight lines on each side, using string or suitable boards (fig. 48). Measure off at right angles and at regular intervals, making a note of the size of the intervals on your field sketch, as well as the actual measurements.

Columns may also be encountered on buildings. As it can never be

Fig. 47

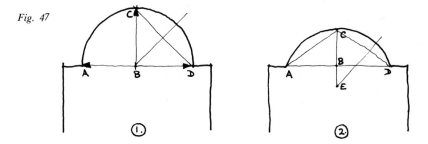

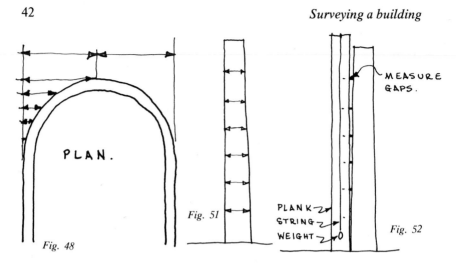

Fig. 48

PLAN.

Fig. 51

MEASURE GAPS.

PLANK
STRING
WEIGHT

Fig. 52

assumed that they are equally spaced, there are two ways of measuring their positions for a plan. Take running dimensions to the centres of the columns, estimating the centres (fig. 49). Measuring the circumference of the column will enable you to work out the actual diameter of the column (diameter = circumference divided by 3.142). Alternatively, take running dimensions from face to face of each column to determine their diameter and spacing (fig. 50).

Some columns taper towards the top (the tapering is called the entasis), and this can also be measured. One way, if a ladder is available, is to make chalk marks at regular intervals and measure the girth of the column at each point (fig. 51). Another way is to use a straight edge, such as a plank, and a plumb bob. Put marks at regular intervals down the plank, and fix the string in the centre of one end. Raise the plank to vertical using the plumb bob as a guide (fig. 52). Estimate the gaps (offsets) at each of the marks. This is less accurate than the first method but can be used where no ladder is available.

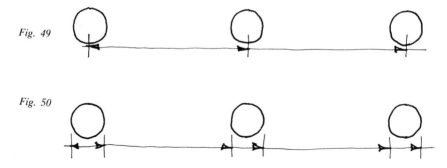

Fig. 49

Fig. 50

(b) Buildings in a state of ruin

There are many stages of ruin. Some ruins are no more than the lower few feet of the ground-floor walls. If the building has collapsed on itself, you will have to be content with measuring what you can reach. At other ruins fallen stone has sometimes been carted away for other buildings, and you can carry out as full a survey, both inside and out, as there are details left to measure. This should at least produce a detailed plan.

Other buildings are in a state of collapse, indicated by fallen slates or cracks in the walls. The only safe procedure with such a building is a survey externally, and even then only with someone watching the roof for loose slates. Take no unnecessary risks — no survey is worth an accident. It would be sensible to wear a helmet.

(c) Tall buildings

The general procedure outlined for (a) still applies, with a few alterations. Start the survey by measuring the building externally at the ground floor. Any thought of hanging the plumb bobs from top-floor windows should be quickly abandoned — it is dangerous to lean out trying to stretch your tape that little bit further. Nor is it much use estimating where upper-floor windows come and making chalk marks — the higher the building, the less accurate such estimates are likely to be. Be content, then, with an external survey of the ground floor only, but make an internal survey of each floor.

The internal survey can proceed as usual but there are two important check measurements to make. One is the vertical height, which can be done in the stairwell using the method described on page 35. This will give the only opportunity of relating all the floors to one another. The second is more a matter of organisation: start the measurement of each floor from a point common to all floors. The points used should be vertically one above the other, such as the same corner of the building or of the lift shaft. This will give a common start to the sketches of each floor and make it easier to trace mistakes should they creep in at some point.

Lastly photograph the elevations of the building from all sides, or from as many as you can. Tall buildings tend to be surrounded by other tall buildings, and you will have to do the best you can. The photographs will enable you to draw in quite accurately details you could not reach to measure. A drawing of an internal section of the building will contain information about window spacing. You will have to guess the thickness of the walls, based on the part you are able to measure from within the building and an estimate of the rest.

(d) Other tall structures

Chimney stacks, railway viaducts and similar structures present problems, and you will have to pick out from this chapter whatever techniques you can bring to bear. Some of these objects are so hemmed in that you can only measure a few surfaces, and the rest of the survey has to be done at a distance.

The height of tall stairless structures is often measured with a theodolite, which measures the angle between the ground and the top of the pile (the angle of inclination) very accurately. If you can borrow a theodolite, and with it someone who knows how to use it, that is clearly the best way. An approximate method of finding the height is to use a clinometer (fig. 53), which is a sighting stick mounted against a protractor. These can be bought cheaply or made.

To measure the height of, for example, a mill chimney, first set up the clinometer so that it is horizontal, using the spirit level (fig. 54). Measure AB and mark it on the field sketch. Sight the clinometer on to the centre of the top of the chimney, and measure the angle of inclination. Use trigonometry tables at home, or in the library, to find the dimension EF. The equation is: EF = AB x tan Θ. Add on AE and you have the height of the chimney. How much else you will be able to measure will depend on the situation of the chimney. Often these emerge from boiler-house roofs, making it almost impossible to attempt to measure the circumference. Others are free-standing, which is easier. Chimneys built before about 1840 were mostly square, and their bases often formed a corner of the boiler-house, which makes these the easiest to measure accurately. There is often nothing you can do to determine the taper of the chimney unless it is free-standing. In that case, measure the lower six feet to find the entasis, as shown on page 42. If the taper of the chimney is constant (it usually is) the rest can be calculated. Where no measurement is possible, take a photograph from as far back as you can and use this to estimate the taper when you make the final drawing, remembering to allow for some distortion in the photograph which will tend to exaggerate the taper. The nearer you were to the chimney when you took the photograph, the worse the exaggeration will be.

The same techniques can be used for railway bridges and other tall structures that you cannot walk on. Another method can be used for road bridges where you have right of access. Weight the tape with something heavy so that it will not swing in the wind, and lower it at regular intervals along the bridge (fig. 55). Sight off the measurements, if necessary using a pair of binoculars. To measure a rising parapet, use a straight edge and spirit level as described on page 22 for making spot levels.

You are likely to find many situations not exactly catered for in this chapter, and you will have to pick methods from here and there to suit the job in hand. One of the excitements of surveying is the continual challenge of the unexpected. No two structures are quite alike (that is why they need

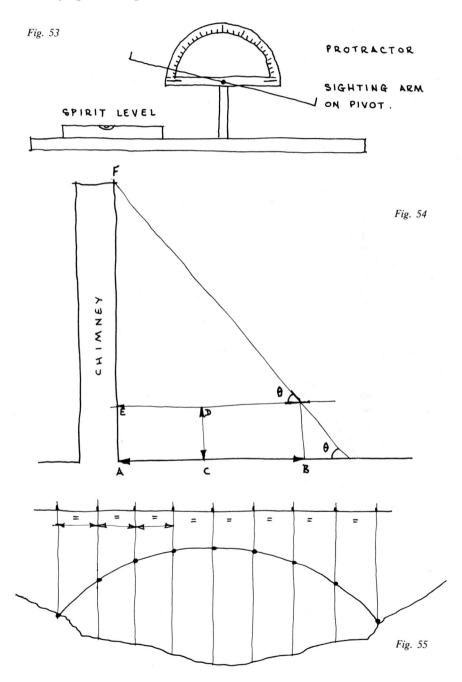

Fig. 53

PROTRACTOR

SIGHTING ARM
ON PIVOT.

SPIRIT LEVEL

Fig. 54

F

CHIMNEY

θ

θ

E   D   B

A   C   B

Fig. 55

to be surveyed) but each problem solved gives increasing experience and confidence to tackle the next. While it is perfectly possible for a person to carry out a survey on his own, you may well find that you learn faster, achieve a more satisfying standard and have far more fun by making up a group of three or four. By doing so, you also bring new ideas to the survey, for the experience and knowledge of the others may well be different from your own. It is therefore more likely that nothing will be missed. You might almost recruit your group for their different skills — photography, sketching, final drawing, making hot soup on the site and a dozen other skills can all come in useful. Jacks of all trades can make a presentable survey; masters of some trades working together should be able to make an even better one.

Drawing up the survey

The equipment you will need for the drawing is a tee square, set square, pair of compasses, some sharp pencils and paper. Paper comes in standard sizes and you need to use the size that is most convenient for what you are doing. In part, this will be determined by the scale you are going to use. A convenient scale for small buildings is ½in or ¼in to a foot, ⅛in to a foot for medium-sized buildings and ₁₆in to one foot for large ones (1:20, 1:50, 1:100, 1:200).

You need a base line to work from, so start by drawing one of the longer walls of the building along the length of the paper. Draw lightly in pencil at this stage of the drawing. From this base, construct the corners of the room, floor or building with compasses, using the tied dimensions on the field sketch. (Do not use the set square for this as you may be putting right angles in that should not be there; very few buildings are absolutely square, no matter how square they appear to be.) Gradually the plan of the building will appear as each detail is plotted from the measurements taken. When all the pencil lines are done, go over the lines of the building with pens and indian ink, and then rub out the construction lines. If you want the drawing to have some character rather than being a soulless mechanical thing, do all the inking freehand. Slightly thicken the lines for the outline of the building and for all door and window openings. Set squares can be used for elevations, for it can be assumed that the great majority of buildings stand vertically. It is helpful if you include a section of the building in the set of drawings you are preparing. This gives a visual report of the wall and floor thickness, etc. To draw a section, make a straight line through a plan and build up the drawing from all the dimensions you have already collected. Indicate on one of the finished plans where the section is by drawing a straight line through it and putting A and B at each end.

Using the photographs, field sketches and your own recollections, try to

CONSTRUCT LINES IN PENCIL
AND COMPLETE IN INK,
FREEHAND, TO GIVE CHARACTER.

Fig. 56

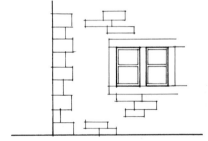

TEE SQUARE & SET SQUARE
PRESENTATION IS BEST KEPT
FOR MECHANICAL & CIVIL
ENGINEERING SUBJECTS.

Fig. 57

match the character of the building in the way you draw the lines. The early nineteenth-century house drawn in fig. 56 was drawn freehand and looks what it is. The same cottage drawn with rulers makes it appear a rather unhappy twentieth-century concoction (fig. 57).

The drawing should now be about finished and is the proper outcome of the hard work in making the survey. You will find that you will be more pleased with the drawing the sooner you do it after the survey, while it is fresh in your mind. Putting it off can too often lead to the plum-stone chant of this year, next year, sometime. . .

Some accepted drawing symbols are shown in fig. 58. An example of a finished drawing can be seen on pages 2 and 3.

Fig. 58

DRAWING SYMBOLS IN COMMON USE.

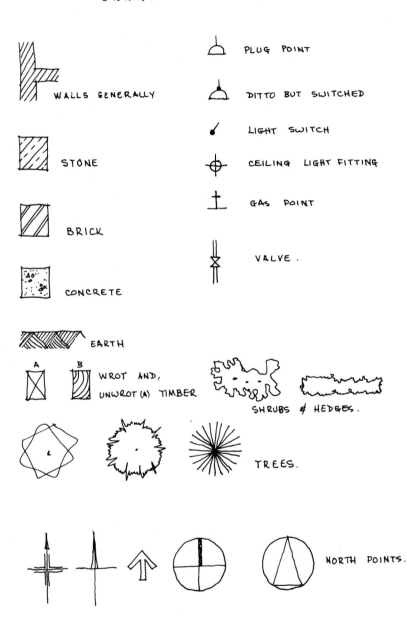

WALLS GENERALLY

STONE

BRICK

CONCRETE

EARTH

WROT AND,
UNWROT (A) TIMBER

PLUG POINT

DITTO BUT SWITCHED

LIGHT SWITCH

CEILING LIGHT FITTING

GAS POINT

VALVE.

SHRUBS & HEDGES.

TREES.

NORTH POINTS.

Check list

1. Sketch all rooms and note wall thicknesses, position of stairs, door and window openings. Label clearly what each sketch is and its location, e.g. FF (first-floor) storeroom.
2. Draw on dimension lines and arrows.
3. Commence measuring in a clockwise direction.
4. Mark in and measure cross ties.
5. Mark in heights — floor to ceiling, sills, window and door heads.
6. Measure vertical tie in stairwell. Measure stairs.
7. Make separate sketches of construction details — floorboards and joists, material of walls, and small details, gas, water and electrical points.
8. Measure and sketch roof trusses, centres and details of construction methods.
9. Sketch and measure external dimensions in anticlockwise direction.
10. Measure or estimate vertical heights, and note numbers of courses of bricks, slates, etc.
11. Photograph each elevation and any special details, including cracks, to give information of character of building.
12. Measure special profiles and mouldings.
13. Note, if desired, layout of drains and nature of foundations.
14. Make a final drawing.

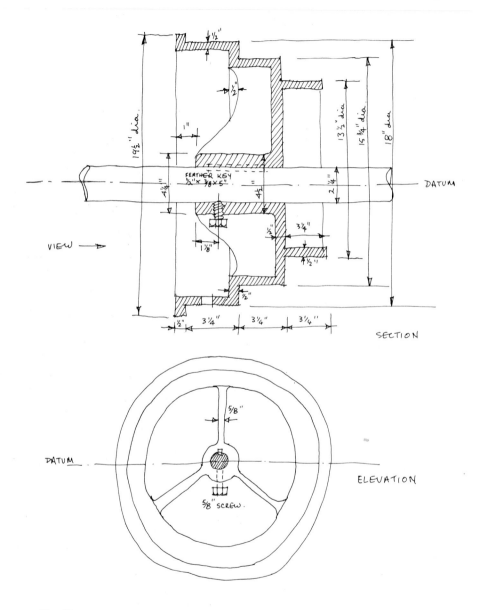

Fig. 59

3. Surveying machinery

A number of sites have remains of machinery which is quite as important as the land and buildings since it is directly connected with the use of the site. It may be a windlass over a well, the shafts and pulleys of line shafting or the bed of an engine. The machines are seldom intact — indeed in most cases they have either been smashed, removed for scrap or even for further use. Usually, then, it is just a few bits and pieces that are left, which may do no more than indicate where machines stood in relation to the rest of the site. This chapter is concerned just with this kind of problem. A reader who wishes to know more about machine surveying is referred to the longer books in the bibliography.

The first steps in surveying the mechanical remains should already have been taken. A well head, for instance, will have been included in the land survey, and its position shown on the finished site plan. Equally, the position of anything standing on floors or fixed to walls or pillars will have been incorporated in the building survey and will also appear in the drawings. All that remains is to take those measurements that will allow you to make detailed sketches to add to the other plans and drawings of the land and buildings.

The first thing to do is to take a close look at the object and select a datum line. This serves the same purpose as the line ED in fig. 1 on page 9 — to provide a starting point for making the final drawing. Sample datum lines are the centre of a shaft or axle, the bed of the machine or some convenient reference point. Next make a rough sketch of the machine. This must be large enough to allow for marking in many fine measurements clearly so that there will be no confusion later in deciding what figures relate to what thicknesses.

After that it is a straightforward matter of measuring every surface, thickness, diameter and so forth. There are no short cuts in this at all, but a patient and methodical approach, a little practice and a dash of common sense will result in a rough sketch with all the dimensions necessary for reconstructing the machine as a finished drawing (fig. 59). Nearly all measurements can be made with a steel rule quite conveniently, as machined surfaces are mostly flat and angles are normally right angles. Be on your guard for those that are not. The most common difficulty is measuring the diameter of pipes, rods and pulleys. A quick but accurate way is to have with you a short length of string with weights on each end.

Hang this over the pipe and measure the distance between the two ends of string, which will be the same as the diameter of the pipe.

Figs. 59 and 60 show that mechanical remains present few new problems to those met in surveying land and buildings because the basic skills of all surveying stem from working out a methodical way of doing things so that no dimension or detail is overlooked.

When you have taken the measurements, make a visual inspection of the machine and make a note of any information that does not appear in the drawing. The check list that follows will give you some idea of the things to look for, but you will need to adapt it freely because there were even greater variations between the products of machine-makers than those of house-builders.

Check list
1. Inspect machine and decide on a datum line.
2. Make a rough sketch and proceed to mark in dimensions.
3. What is it made of — cast iron, wood, other metals?
4. Is it painted or not? If painted, make a note of colours and anything that seems to be unusual in the way of decoration, such as lining.
5. Does the machine look as if it was made by a specialist firm or 'home-made'?
6. Is there a maker's plate, number, patent number or other writing? If a maker's plate has been removed, what shape was it? Some plates had distinctive shapes.
7. Are there any marks to indicate the use of the machine — pulley belts leave wear marks if they rub against pillars? Is there grease on the floor or walls around a machine bed? Is there charcoal on the floor? Are there unnatural wear marks or depressions anywhere? The less there is of the machine, unless you know what the building was used for, the more you have to search for information around the remains.

It is tempting to think that if you have surveyed a building and its site which is marked on an old Ordnance Survey map as a mill, that there is no real need to bother about the fussy details of machinery. The thought usually springs from the fact that you are always moving about while measuring a building and therefore seem to be making rapid progress, whereas with a machine you are standing on the same spot for a long time. Do not be tempted to ignore the detailed work, for it may well reveal the real purpose of the mill and make sense of details in the architecture that otherwise seem inexplicable. There were, after all, many kinds of mill — for grinding corn, snuff and logwoods, for spinning, weaving or finishing

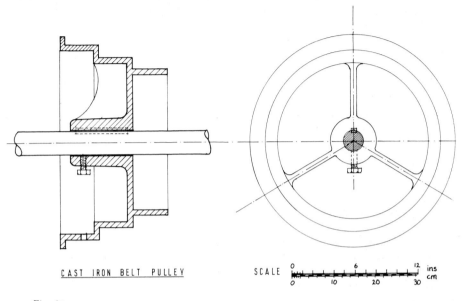

CAST IRON BELT PULLEY   SCALE

Fig. 60

wool, cotton, linen, silk or other fibres, for rolling iron and drawing wire, and many more. Mills were frequently used for different processes over a period of time or in various parts, and yet mill buildings often look alike. A patient and methodical record of mechanical remains may well provide the clues to explain the use of buildings or the actual way in which a particular process was done. The final drawings that we are encouraging you to make are intended to be a permanent record. You may not fully understand all the details that you record but someone who has seen similar sites in other areas may well be able to interpret what you have recorded, provided all the details are there. The more thorough your survey, the greater value it will have when the site has been swept away. A survey that only shows what the surveyor thought important or can understand is often more irritating than informative.

4. Film and tape

The other chapters in this book are concerned with the procedures for making surveys of land, buildings and machinery, and making the final drawings. Mention has been made several times of the need for photographs, both taking them and using them later. This chapter will pursue the matter a little further but will also go beyond the confines of a strict survey for reasons that will emerge.

Several excellent books are available that explain photographic technique, and nothing need be added to those. The question is rather how much can be done with equipment readily available, and what part this plays in the survey. This immediately raises the question of black and white versus colour, but first let us look at the camera.

We have tended to assume in writing this book that there is an urgent reason for undertaking the survey at this particular time — the building is about to be demolished or the site developed. In these circumstances, you will have to use whatever camera you have or can borrow. This need not be a fancy one, with built-in this and automatic that. Some of the most useful survey photographs are taken with an elementary box camera. This is quite adequate for views of elevations and for most details; only the small details may need a camera that can be focused more finely. Some box cameras can even be used with flash guns and so are serviceable for photographs inside the house. The main drawback with box cameras is that they need a lot of light — best results are obtained in bright sunlight, and edges tend to be blurred as the light grows less. On the whole box cameras can take a better photograph in poor light with black and white film than colour. While a box camera can give adequate results, if a more versatile camera is available, use it.

As with most things, it is better to use one that you are used to than borrow one you have not tried before. You will not know whether you are doing the right thing until the prints come back, and that may be too late. If you are going to borrow the camera, why not get its owner interested too and let him take the pictures for you?

The photographs are wanted for three main reasons. In the first place, they act as a safety net, recording any details you may have overlooked in the survey as well as supplying information you may not have been able to measure because it was out of reach. It is helpful if there is some indication of scale in these photographs. Use a ranging rod if you have one available, or failing that, anything of a standard size. Milk bottles, Ordnance Survey maps, people and animals all give the approximate scale that is needed. Photographs of details need a more accurate indication of size, such as a centimetre scale. The second reason for having photographs is that you may need them when you make the final drawing, to give an indication of the character of the building, or other details. Finally, some photographs are useful to complete a full report on the subject of the survey. The drawings will have all the relevant measurements and will give a clear and accurate portrayal of whatever it is; the photographs show you what it looked like, how much wear it had, what state of repair it was in and similar information which could only be included in a survey if it were written out in full. Exterior pictures, for example, give a good idea of the nature of the land — what kind of plants are growing, do they grow evenly or are some patches stunted by dumps of chemical waste, to what extent do the tall mill walls keep the light from the houses on the other side of the road. Some of this information may be contained in a detailed survey, or it may be mentioned in an accompanying report. The advantage of a photograph is that it provides visual evidence of these matters, and therefore is a valuable contribution to the total report. For if you have gone to so much trouble to make the survey and the final drawings, it is surely worth going a little further to make out a report which contains all the information you can gather about the site.

More will be said about that in the final chapter but first there is more to say about photographs. You may by now have realised that there is a good reason for using black and white film in preference to colour. There are several, in fact. One is that it is cheaper. This means that you can afford to take many more photographs than if you were using colour, and, if you look at the negatives carefully, you need only order prints of the ones that tell the most. The more you take, the less chance there is of omitting details.

Another reason for using black and white is that it is relatively stable compared to colour. A colour photograph soon starts to fade when exposed to light, rather like the early black and white pictures. Modern processing has eliminated most of the fading tendencies from black and white photographs, and these will now last several decades with reasonable care. More important, if the negative is properly protected, it will stay as clear as it ever was for a very long time, so that other prints can be taken from it. Colour negatives do not at present have anything like such a long life.

Yet another reason for keeping to black and white film is that you need the photographs to work from. You may have to look through a batch of pictures quickly, hunting for some detail. You will want the prints as large as possible for this, to make it easier to spot what you are seeking. Half plate or whole plate enlargements of black and white photographs are expensive enough, but the cost of those sizes in colour would be out of the question. The alternative, of course, is colour slides. You could look at these on a hand viewer but it will do your eyes no good looking for details on such a small picture. Alternatively you might project them on to a screen. This would certainly give you a good size, but would it be much help? Best projection is obtained in a darkened room; the best drawings are done in a well-lit room. Somehow you would have to reconcile the two. You will also find it much harder to translate details from a screen than from a photograph beside your drawing.

One last point, though not of itself sufficient to influence your decision, is the possibility of publication. The majority of surveys are never published in book form, but suppose you did have one published. If the full report contains photographs, then they should be published too. This may well be possible if black and white prints are available, but few publishers would consider colour photographs because of the high cost.

What, then, is the use of colour film in a survey? For the survey itself, very little. Colour negatives and prints are so expensive and prone to fade that they are best ignored for these purposes. Colour transparencies (slides) can help to supplement the other photographs particularly if you find it hard to visualise the site from black and white photographs and want to refresh your mind before making the drawings. Clearly these could be very helpful if there has been a gap between the survey and the drawing. Apart from this, slides only come into their own to remind yourself in later months of what an enjoyable time you had making the survey, or you can use them to talk about the site to groups of people. If you are likely to use the slides a lot in this way, as a teacher might, you would be well advised to take two of everything. This again is to reduce the effect of exposure to light. Use one slide for your lecturing and put the other away in a dry well ventilated place. This can be used for making a copy later on, should it ever become necessary. You could just take one and have a copy made from that, but you will probably find it cheaper and more satisfactory to take two in the first place.

Moving further up the scale of costs brings us to films. The first thing to stress is that they are not necessary to making a first-class survey — the quality of the measuring and the final drawing determine that. You should not feel that you cannot do a good survey because you cannot make a film. They are not necessary. They are only mentioned because there are some situations where they can add something to a report that would be hard to put there in any other way. Examples of such situations are that the waterwheel is set in motion or the derelict lock gates moved. A film just of buildings can be helpful by showing one building in relation to the rest of the area, so creating the atmosphere of the area. As 8mm cameras seem more widely dispersed than they used to be, it is worth considering whether the site has anything to offer that can be filmed. Films are expensive, though, and can never replace a careful and well-drawn survey for future students seeking to learn about the site concerned.

Tape recordings fall into much the same category as films, though they are not quite so expensive. The noise of the machine running can be very informative as well as fascinating. Far more use can be the memories of the people who worked the machine, lived in the building, cut the grindstones, and so forth. Dialect has been well recorded in most parts of Britain but much useful work remains to be done collecting reminiscences and folklore. Much of this is to do with work and home and is therefore appropriate to this chapter. There are many people in most communities who can not only remember much from their own childhood days but also what their grandparents told them of theirs. It would be as foolhardy to believe everything that is recorded as to attempt to make a model from some of the drawings made in the middle ages. However, we would know little about some industrial processes if it were not for these drawings, which can be modified and interpreted from the scattered references in contemporary writings and from the results of archaeological excavations. Taped reminiscences can also be modified and more so when more are available for comparison. Perhaps a responsible body will encourage the collection of such memories and compile a national catalogue so that comparisons can be made. Inevitably each year that passes reduces the choice of people with recollections of the distant past. The more industrial archaeologists and others include this kind of survey work in their activities, the greater the variety of information available when a national body is formed to catalogue them all.

As an illustration of what can be done in this way, the authors knew a man living near them of about ninety years old. He had a clear memory and could remember very well crouching under a hand loom at the age of four, warning his mother who was weaving when the threads broke and fell down. He could describe the steam engine at the mill in which he worked for more than sixty years, and how this was coupled to the 30ft waterwheel on a common shaft. (The wheel, engine and unique regulator have all been broken up, and there is little information about them in the firm's papers.) More than that, he also remembers many things that his grandfather told him, about life in the village when he was young, and he was born in 1815. His memories have been recorded. You will not always find people with such remarkable memories but you can expect to find many who have worked at a trade or on a machine now obsolete. Such people are usually delighted to be asked to talk about what they did and how they did it, once they have overcome their initial shyness. You can help them to do this if they have not been recorded before by letting the recorder run for a few minutes while you have a chat about almost anything and then play it back. After a laugh about how funny everyone sounds, you can then ask the questions that will lead to the memories you would like to have on tape.

The point we have now reached is somewhat removed from the use of tape recordings to supplement a building survey. We have gone on from that to encourage you to make use of a tape recorder to make a different survey — a survey of people's memories of the work they did, how they used the tools, what they were paid compared to the prices they had to pay and so forth. This kind of record is being made in a few places but a vast amount more could be done. A national catalogue of what already exists, including the significant collections of such bodies as the BBC, would show where the gaps are so that they could be filled while there is still time. There are so many semi-official cataloguing organisations in the country already that we hesitate to suggest the need for another. However, this kind of survey would seem to be desirable and, if there is not to be much unnecessary duplication of effort and the inevitable gaps, someone needs to find out what has already been done and make the information available in all reference libraries. Industrial archaeology is about people. Surveys of the buildings the people lived and worked in are a valuable contribution to understanding what their lives were like; their views on tape add flesh to what could otherwise become a very dry and lifeless exercise.

5. Completing the report

It is at this stage that many surveys come unstuck, and the usual reason is that the survey is never completed. A common pattern in the way surveys are attempted is that the ringleader and his friends hear a place is threatened with demolition, rush over and survey it and talk of doing the drawings when the photographs come back, or in the winter. By then, other matters seem more urgent, and the site drawings gather dust. This is a thousand pities, since the survey is of no use to anyone unless it is completed — it might as well not have been done. Since the point of making this kind of survey at all is to record something of importance before it disappears, it is clearly vital that the enthusiasm which started the group on doing the site work must carry them over to making the finished report.

The first stage in completing the report is to finish the drawings. These need to be done on large sheets of good-quality paper, such as cartridge paper. This will enable them to stand up to the wear and tear of being handled. Alternatively, if the drawings are likely to be copied to any great extent, heavy tracing paper can be used. This is not quite so hard wearing as cartridge paper and cracks easily, but it is easier to copy. Paper comes in standard sizes, and both imperial and metric sizes are available. The approximate equivalents are:

imperial size	*metric size*
half imperial (13½ x 19in)	A3 (297 x 420mm)
imperial (19 x 27in)	A2 (420 x 594mm)
double elephant (27 x 36in)	A1 (594 x 841mm)

In addition, both cartridge paper and tracing paper can be obtained in rolls about 30in or 750mm wide. Whatever the drawings are done on, the paper needs to be securely fastened down to a drawing board or some other flat surface with clips or adhesive tape before you start to draw: a sheet that can be pulled along with a pencil is a nuisance.

The equipment needed for drawing has been referred to already: a pair of compasses, set square, protractor, long ruler, rubber and a hard pencil (grade H or HH). A hard pencil will draw a fine line without repeated sharpening. The ruler must be in whatever units you have used for the survey, whether metric or imperial. You also need a pen and indian ink for inking in the final drawing (black felt-tip pen tends to fade). The paper and other equipment can normally be obtained from good stationers or suppliers of artist's materials.

Mention has been made in the appropriate places of the methods of making the drawings, and little needs to be added. It is important to relate all the drawings to each other. This can be done in the case of small buildings by putting all the drawings in sequence on a large sheet. Each floor should be correctly labelled, elevations related to the appropriate side of the ground-floor plan, and the whole group to the site plan. Larger buildings and complex sites will spread over several sheets, which makes it all the more important to relate them to each other either by a sequence of letters, a detailed caption to each drawing or by a full list of the series of numbered drawings. The thought at the back of your mind should be that someone who had never seen the site could imagine it from your set of drawings.

The drawings should be accompanied by written notes on anything that warrants it. Some ideas for this were included in the topographical survey and elsewhere. Other matters deserving full comment are details of materials used for constructing buildings and the methods of construction. Any dates you have found either on the site or from later reading should also be included, as well as any other definite historical facts. 'The home of Goldsworthy Gurney from 1820 to 1831' is more helpful than 'King Charles is thought to have slept here once'. However, writing a history of the site is a lengthy and quite separate business, and it is better to jot down the few facts you have rather than delay finishing the survey while you study the history in depth. Include in the report the authors, titles, publishers and dates of publication of any books that contain references to the site, and the names and addresses of anyone who has photographs, notes or any other information.

Finally, the complete collection of drawings, notes, reports and photographs needs to be fastened into a file so that nothing goes astray when it is lent to someone else. A ring binder is a strong protection for such a survey, or a box file if there are slides, film or tape as well.

Now what do you do with it? If you have not already done so, you should tell your local museum, library, industrial archaeology group or local history society what you have compiled. We hope you will have done this before you started the survey anyway, partly to make sure that no one else had surveyed 'your' site already and so avoid unnecessary duplication, and partly to enlist the support and help of others. Now let them know that it is finished. The local library or museum may make a note of the nature of the survey so that they can refer enquiries to you should any arise. They might even wish to copy the survey in order to add it to their reference collection, though they have to watch the costs of such work. A lively industrial archaeology or local history society will wish to comment on the survey in their bulletins to keep members informed of work being done. You or they should also inform the Council for British Archaeology, which started a national catalogue of industrial remains, administered from the University of Bath. Some readers may be reluctant to publicise their work, because they are shy, modest or feel that their work would not be good enough or interesting enough to be seen by anyone else. We hope they will overcome these feelings because the whole point of making the survey, even if it is the first they have done, has been to make a permanent record of a part of the nation's heritage that was threatened with extinction. As we said at the start, it is totally impossible to keep every historic building and site, and even many that can be kept must be altered or adapted to new uses. It is possible, though, for full surveys to be made of everywhere of interest. However, this will not be of much use unless the existence of the surveys is made public. So please tell somebody what you have done — local newspapers are always interested in people who have done something of use to others.

What about publication? This is quite another matter, for printing costs are very high. Survey reports are seldom printed; those that are tend to fall into one of the following groups: either they are about sites or buildings which are visited by large numbers of people, such as industrial museums, or the survey is an outstanding piece of work, or the site is in some way unique — perhaps it is to do with an unusual industry, or an ancient craft, or is connected with some famous person or development. Reports have been printed on topics outside these groups, especially when the *Journal of Industrial Archaeology* was being published, but less frequently. The most likely people to be interested in printing a survey are again the local museum, library, local history or civic society. Only a handful of sites holds much interest for people outside the area, and any publisher has to think hard about where the money will come from; five hundred copies is too small a number to print to achieve any economies but it is a large number to sell. So do not be discouraged if no one seems keen to publish a survey. Indeed it is better to tackle the survey as something that is worth doing and enjoyable in itself, and look on publication as a happy bonus if it happens.

That brings us back to the point at which we started. Britain's industrial past is something that stretches back not just to the industrial revolution but for more than two thousand years. There are many reminders of this past remaining and it is important to record these before they are all swept away. Road programmes, housing estates, factories and reservoirs require the demolition of large numbers of old properties every year. Among them are some which should be surveyed in full. A much greater number are slowly decaying through neglect or are being subtly modernised out of all recognition. These disappear almost without comment. Surveys are needed of these too — it is better to survey a site while it is still in use than wait until it is a ruin and have to guess the significance of some features. It is important also that all kinds of sites should be tackled, not just the fashionable or pretty ones. A 1930s filling station may not have the appeal of a water-powered mill but it is just as much part of industrial history. It is perhaps inevitable that people will think nothing of travelling fifty miles to survey the mill but would not give the 1930s garage down the street a second glance. There are usually far more worthwhile sites close at hand than meet the eye, and it is certainly easier to make an unhurried survey of one of those.

There are other remains of Britain's industrial past which are also threatened with obliteration. Old films, photographs and lantern slides are still being destroyed or sent abroad as antiques, yet they frequently contain valuable information about industrial matters. The films made by firms to train apprentices or attract customers are particularly vulnerable to being thrown away when they are dated, though parts of them should be retained by someone. The English language is full of words and phrases that spring from occupations or processes, but these too are fast disappearing as communities break up and we all try to be what we think the television would have us be. This kind of information also needs to be recorded, and the publication of pamphlets listing the local words and phrases would be valuable to anyone studying history and might prevent some false assumptions. Ultimately what is needed is a dictionary of historical industrial language but a lot of work remains to be done in collecting the basic information.

Britain's industrial past is considerable but many of its remains are fast disappearing. We hope that an even larger number of people will examine the area around them closely, identify and survey sites, complete the survey reports, collect the meanings of local industrial words and phrases, seek out film and so forth — and above all that they will enjoy it.

Bibliography

Fieldwork

Aston, M. and Rowley, T. *Landscape Archaeology.* David & Charles, 1974.

Clendinning, J. and Olliver, J. G. *Principles and Use of Surveying Instruments.* Van Nostrand Reinhold Co, 1950.

Knight, B. H. *Surveying and Levelling for Students.* CR Books, 1968.

Major, J. K. *Fieldwork in Industrial Archaeology.* Batsford, 1975.

Pannell, J. P. M. *The Techniques of Industrial Archaeology.* David & Charles.

Perrott, S. W. *Surveying for Young Engineers.* Chapman & Hall, 1946.

Starmer, G. *Industrial Archaeology of Watermills and Water Power.* Heinemann Educational, 1975.

Taylor, C. *Fieldwork in Medieval Archaeology.* Batsford, 1974.

Drawing

Dove, E. D. and Gaskin, A. T. W. *Modern Engineering Drawing.* Pitman, 1967.

Gill, R. W. *The Thames & Hudson Manual of Rendering in Pen and Ink.* Thames & Hudson, 1973.

Jackson, T. and Bentley, P. *Engineering Hand-sketching and Scale Drawing.* Pitman, n.d.

Mayock, B. *Technical Drawing Book 5 — Engineering Drawing.* Heinemann Educational, 1969.

Pearson, G. *Engineering Drawing.* Oxford, 1960.

Thomson, R. *Basic Draughting Practice.* Nelson, 1969.

Industrial history

Bodey, H. *Discovering Industrial Archaeology and History.* Shire Publications, 1975.

Bodey, H. *Textiles.* Batsford, 1976.

Bodey, H. *Twenty Centuries of British Industry.* David & Charles, 1975.

Campbell, W. A. *The Chemical Industry.* Longmans, 1971.

Derry, T. K. and Williams, T. I. *A Short History of Technology.* Oxford, 1960.

Greaves, W. F. and Carpenter, J. H. *A Short History of Mechanical Engineering.* Longmans, 1969.

Hudson, K. *Building Materials.* Longmans, 1972.

Raistrick, A. *Industrial Archaeology — An Historical Survey.* Paladin, 1972.

Reynolds, J. *Windmills and Watermills.* Hugh Evelyn, 1970.

Simmons, J. *Transport.* Vista, 1962.

Index